MAKE it WORK!
ELECTRICITY

Wendy Baker & Andrew Haslam

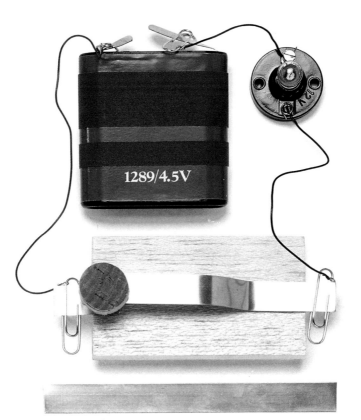

1289/4.5V

written by
Alexandra Parsons

Photography: Jon Barnes
Series Consultant: John Chaldecott
Science Consultant: Graham Peacock
Lecturer in Science Education at
Sheffield City Polytechnic

MAKE it WORK!
Other titles

Body
Building
Dinosaurs
Earth
Flight
Insects
Machines
Photography
Plants
Ships
Sound
Space
Time

First published in Great Britain in 1992 by
Two-Can Publishing Ltd
346 Old Street
London EC1V 9NQ

Copyright © Two-Can Publishing Ltd, 1992
Design © Wendy Baker and Andrew Haslam, 1992

Printed in Hong Kong

Hardback 6 8 10 9 7 5
Paperback 2 4 6 8 10 9 7 5 3

A catalogue record for this book is available from
the British Library

Hardback ISBN: 1 85434 134 0
Paperback ISBN: 1 85434 109 X

Editor: Mike Hirst
Series concept and design: Andrew Haslam and Wendy Baker
Additional design: Belinda Webster
Thanks also to: Albert Baker, Catherine Bee, Tony Ellis,
Elaine Gardner, Nick Hawkins, Claudia Sebire and everyone at Plough Studios.

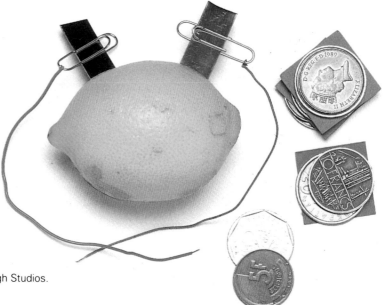

Contents

Being a Scientist   4

Static Electricity   6

Simple Circuits   8

Batteries   10

Lighthouse   12

Complex Circuits   14

Light Bulbs   16

Circuit Game   18

Switches   20

Morse Code   22

Circuit Quiz   24

Magnetism   26

Magnetic Power   28

Magnetic Boats   30

Magnetic Earth   32

Magnetic Fields   34

Electromagnetism   36

Making a Motor   38

Working Motor   40

Electric Train   42

Boosting Power   44

Glossary   46

Index   48

Be Careful! Electricity can be very dangerous. All the activities in this book use batteries as the power source. You should **never** experiment with mains electricity or the plugs and sockets in your home.

Words marked in **bold** are explained in the glossary.

Scientists study the world around them and the way it works. They ask themselves questions and then work out systematic methods for answering those questions. Scientists usually start out with a **theory,** which they test by doing **experiments**. Then they observe and record the results.

Investigating electricity is part of the science of **physics**. Physicists study **energy** and **matter**, and find out how to put the forces of nature to work on our behalf. For instance, physicists discovered how to make **hydroelectricity** and **atomic** power.

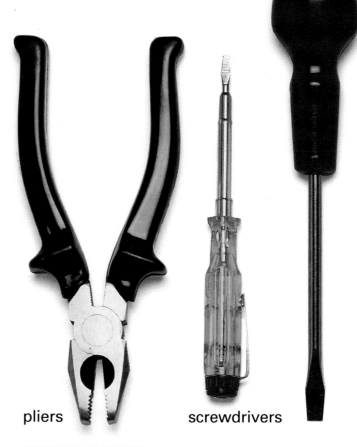

pliers   screwdrivers

MAKE it WORK!

This book is all about electricity. As you do the projects, you will be investigating the science of physics for yourself. It is important to use scientific methods. Draw pictures as accurately as you can, or take photographs. Write down clearly what you have done and observed.

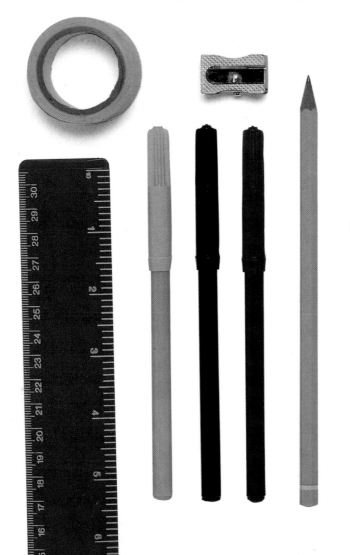

peg and foil bulbholder

LED bulb

small bulbs

crocodile clips

bulbholders

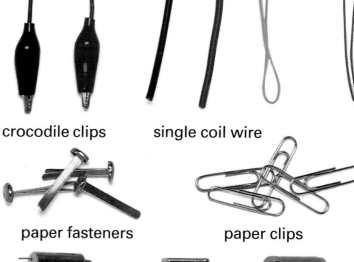

crocodile clips   single coil wire

paper fasteners

paper clips

electric motors

You will need
Notebooks, pens and pencils, tape, ruler, scissors and a protractor. Specialist equipment needed for the projects can be bought from any electrical goods or hobby shop.

Wire cutters and **pliers** Special wire cutters are best. You can use an old pair of scissors, but the wire will make them blunt.

Screwdrivers You will need a small, insulated electrical screwdriver and a larger one for screwing pieces of wood together.

Bulbs and bulbholders Use 6-volt bulbs and matching bulbholders. A square of tin foil held in place by a peg makes a good substitute bulbholder.

Wire Use single-core, plastic-coated wire.

Clips Buy crocodile clips for connecting wires to one another or to batteries. You can also use ordinary paper clips or paper fasteners.

Small electric motors These come from hobby shops in a variety of shapes and sizes. Use 3-volt or 6-volt motors.

Buzzers and **magnets** These are sold in model shops and hardware stores.

Batteries Most of the activities in this book use simple 6-volt or 4.5-volt flat batteries.
Be careful! Never touch car batteries and never plug anything into the mains sockets in your home. Mains electricity and large batteries are extremely dangerous!

horseshoe magnet

battery holder

4.5-volt flat battery

Have you ever noticed that when you brush your hair, it sometimes sticks to the comb? That happens because of **static electricity**.

Every single thing is made up of tiny particles called **atoms**. Normally atoms have no electrical activity, but when two things rub together, like hair and a comb, the outer layer of **electrons** on the atoms of the hair are rubbed off. They stick to the atoms on the comb. When atoms lose electrons we say they become positively charged. When they gain electrons they are negatively charged. Two like charges repel one another – and different charges attract.

Repelling

Rub a balloon against your jumper and ask a friend to rub one too. Tie the balloons to a stick, with the rubbed sides facing each other. Because both balloons have the same charge, they swing away from one another.

MAKE it WORK!

With an **electroscope** you can test for the presence of static electricity.

You will need

bare wire
aluminium foil

a glass jar with a plastic lid
foil from a sweet wrapper
a plastic pen or ruler

Ask an adult to help you push a piece of wire through the lid of a jam jar. Bend up one end and drape a thin piece of foil from a sweet wrapper over it. Crumple a ball of aluminium foil around the other end. Rub a plastic pen with a piece of silk or wool, and then hold it over the foil ball. If the pen is charged, then the sweet wrapper will move.

Attracting

Make some piranha fish like the ones above. Using the graph paper shape as a guide, cut out the fish from a single layer of coloured tissue paper. Place them on a flat surface. Rub a plastic ruler on a piece of silk or wool to get the ruler's electrons moving. Now pass the ruler over the fish and watch them jump up, attracted by the electric charge.

▼ Make some curly tissue-paper snakes and decorate them using stencils or felt-tip pens. (Be careful because tissue paper tears easily.) Pass a charged ruler over them – and watch them wiggle and wriggle!

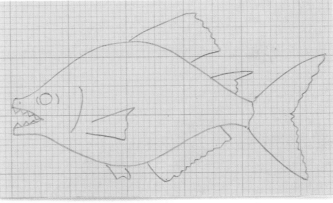

Lightning is caused by static electricity. When water droplets and ice particles in a cloud are rubbed together by air currents and strong winds, they produce an electric charge. When this charge gets very strong, a mighty stream of electrons leaps down through the skies towards the ground as lightning.

Static electricity itself is not very useful to us – we have to harness electricity before it can be used. The power that we actually use in our homes is called **current electricity** and is made up of millions of moving electrons.

An electric current is formed when the electrons in a substance, such as a piece of wire, are all made to move in the same direction. To provide us with electrical energy, the electrons must flow in an uninterrupted loop, called an electrical **circuit**.

—|⊢— battery

▲ All the projects and activities in this book use batteries as the source of power.

bulbs   bulbholder

▲ Scientists and electrical engineers use special symbols when they are drawing or designing a circuit. This symbol is a bulb inside a bulbholder. A drawing of a whole circuit is called a circuit diagram.

MAKE it WORK!
Try making a simple circuit for yourself. It is very easy, but you must check carefully that all your connections are properly made.

You will need
a battery
wire
a bulb and bulbholder
paper clips/crocodile clips
a collection of household objects

1 Cut two pieces of wire about 15cm (6 in) long. Strip the plastic coating from the ends of the wire, without breaking the wire itself.

2 Attach one piece of wire under each of the connecting screws on your bulbholder.

3 Attach the other ends of the wires to the battery **terminals**. If all your connections are made properly, the bulb should light up.

crocodile clips

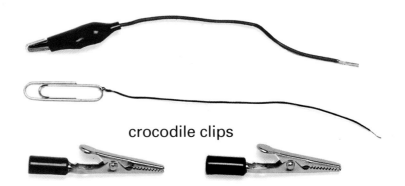

*Anything that electricity can flow through, such as metal, is called a **conductor**. Materials, such as plastic, rubber and glass, that do not allow electricity to pass through them are called **insulators**. Electrical circuits make use of both conductors and insulators. The conductors, in this case the metal in the wires, allow the current to flow around the circuit. The insulators, such as the plastic around the wires and on bulbholders, stop the current from passing into any metal objects that the circuit is touching.*

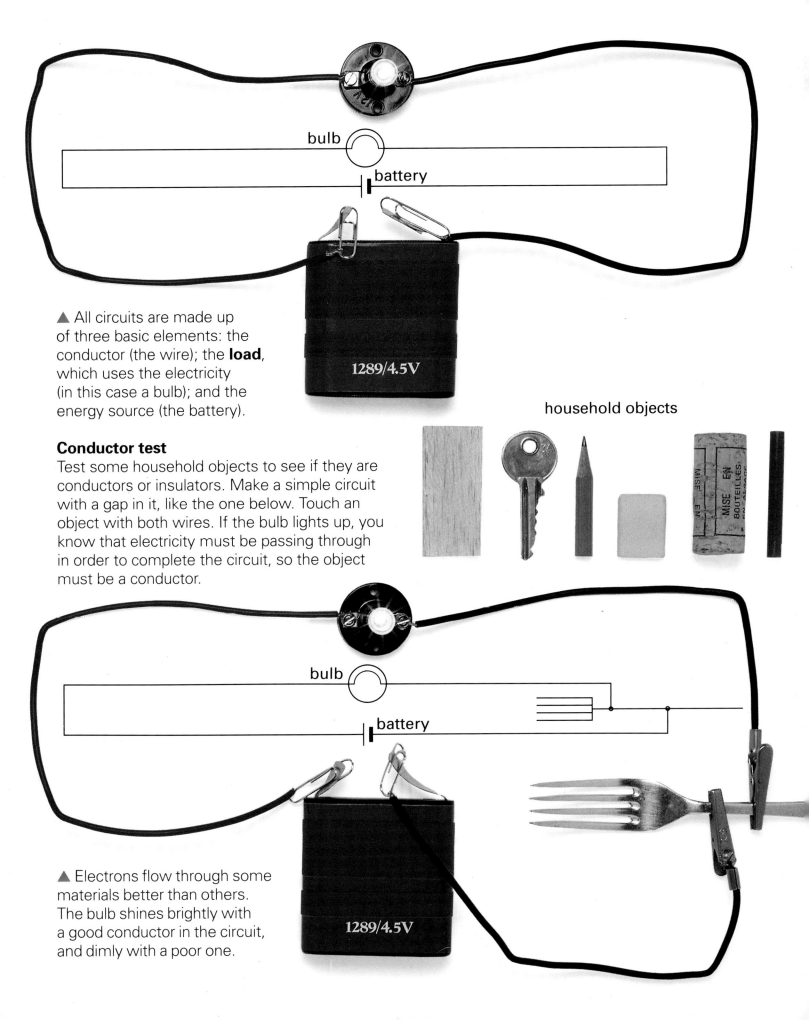

bulb

battery

1289/4.5V

▲ All circuits are made up
of three basic elements: the
conductor (the wire); the **load**,
which uses the electricity
(in this case a bulb); and the
energy source (the battery).

Conductor test
Test some household objects to see if they are
conductors or insulators. Make a simple circuit
with a gap in it, like the one below. Touch an
object with both wires. If the bulb lights up, you
know that electricity must be passing through
in order to complete the circuit, so the object
must be a conductor.

household objects

bulb

battery

1289/4.5V

▲ Electrons flow through some
materials better than others.
The bulb shines brightly with
a good conductor in the circuit,
and dimly with a poor one.

10 Batteries

Batteries produce electricity from chemical energy. Usually, two metals, called electrodes, are placed in an **acid** solution called an **electrolyte**. A chemical reaction takes place and creates electric power.

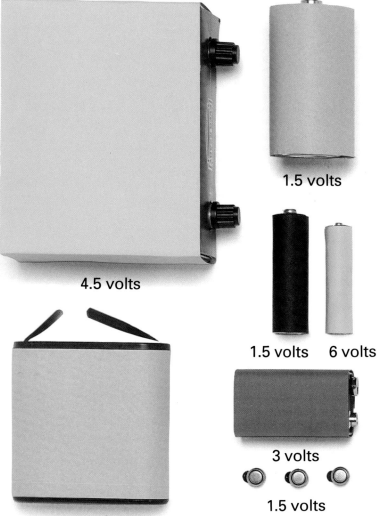

4.5 volts

1.5 volts

1.5 volts   6 volts

3 volts

3 volts

1.5 volts

Positive and negative

An electric current needs a destination in order to keep it moving. In a battery, two metals, zinc and copper, are used to make a current. When they are put into acid, negative electrons move from the copper to the zinc through the liquid. From the zinc, they move back down the wire to the copper, causing an electric flow.

MAKE it WORK!

There are many different kinds of battery. Wet batteries have metal plates in a liquid acid. In dry batteries, a chemical paste separates a carbon rod from the zinc case. Other batteries contain the metals nickel and cadmium, and an **alkaline** substance instead of an acid.

To make a wet battery you will need

a glass jar   white vinegar
wire   crocodile clips/paper clips
a strip of zinc   a piece of copper pipe
a light emitting diode (LED)

1 Put the strips of metal in the jar and fill it with vinegar. (Vinegar is a kind of acid.)

2 Attach clips and wires as shown, and the bulb will light up. However, LEDs only work when wired up the right way round. If yours doesn't light first time, reverse the connections.

The first battery was invented by an Italian count, Alessandro Volta, in the 1790s. It used silver and zinc discs, rather like our coin battery.

To make a battery tester you will need

balsa wood

insulated copper wire

screws and washers

card

a compass

wire and clips

1 Wrap the compass and the backing card in copper wire, attaching the ends to screws on the wooden base as shown.

2 To test a battery, clip wires from the battery terminals to the screws. The compass needle will move. Try this experiment with a brand new battery and a battery that has been used a lot. Can you notice a difference?

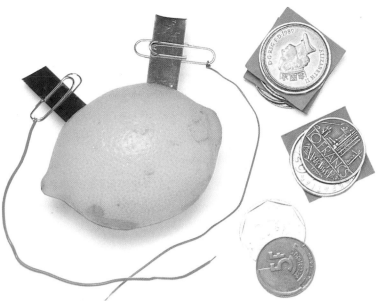

▲ **Mini-batteries** You can make a coin battery using silver and copper coins. Pair up the coins, and separate each pair with a square of blotting paper soaked in salty water. Attach a wire to the bottom coin of the pile, and a wire to the top coin. Don't let the wires touch each other, but clip them to your battery tester and see what happens. The current won't be very strong, but the tester should make some reaction.

You can also make a low-voltage mini-battery by pushing copper and zinc strips into a lemon.

Electricity has many uses – in homes, factories and schools. It is produced in power stations by burning coal or oil fuels to power electricity **generators**. It can also be produced from nuclear fuel or in hydroelectric turbines.

MAKE it WORK!

Lighthouses were among the first users of electric power. Put your circuit-building know-how to good use and make a battery-operated mini-lighthouse for your bedroom.

You will need

thin card	glue and sticky tape
a craft knife/scissors	wire
a bulb and bulbholder	a battery
crocodile clips/paper clips	

1 Make a round tube from a piece of white card and decorate it with red stripes. You could also use the cardboard tube from a toilet roll and cover it with white paper.

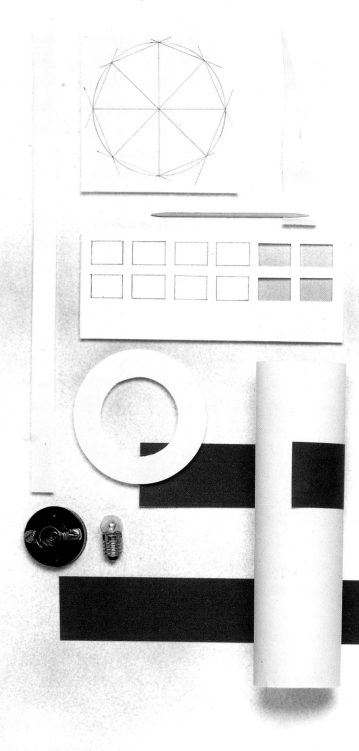

2 Cut out a circle of card for the balcony, make a hole in the centre, and glue or tape it to the top of the tube. Glue a strip of card around the edge of the balcony to make the rail.

3 Attach two long wires to a bulbholder and tape the bulbholder into place at the top of the tube. Push the wires down through the tube and out at the bottom.

4 Take a strip of card to make the windows at the top of the lighthouse. Cut out small squares with a craft knife or scissors, using the picture on the left as a guide. Then bend the card round to make a cylinder shape and glue it in position on the balcony.

5 To make the roof, draw a circle. Make two cuts close together from the rim to the centre and cut out a small segment. Fold and glue the circle to form a cone. Make a flag from paper and a cocktail stick.

6 Attach the ends of the wires to a battery and the bulb in your lighthouse will light up.

Scientists measure electricity with two separate units called volts and watts. Volts measure electrical force, the amount of power produced by a source of electricity, such as a battery. Watts measure the electrical power at the point where it is actually used – in an electric fire or bulb, for instance.

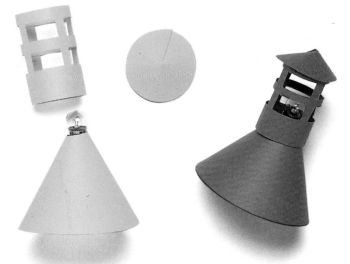

The first people to build lighthouses were probably the Ancient Egyptians. They began by lighting bonfires on hilltops to guide their ships. During the third century BC, they built the tallest ever lighthouse, the Pharos of Alexandria, which was over 122 m (400 ft) high.

▲ To hide the battery that operates your lighthouse, make a 'rock' out of pieces of old cardboard, stuck together in a jagged shape and painted. Around the lighthouse, put a series of buoys like those on the next page.

In an electrical circuit, all the parts must be joined up to one another, so that the current can flow. There are basically two ways of wiring a circuit with more than one **component** (or part) – in series or in parallel.

MAKE it WORK!

In a series circuit, the electric current flows along a single path, going through each of the components in turn. If one component is removed, or breaks (when the filament in a bulb burns out, for instance), all the other components will stop working too.

In a parallel circuit, each of the components is connected to the battery on its own branch of the main circuit. Even if one of the bulbs in a parallel circuit burns out, the other bulbs will continue to shine, because their own branches of the circuit remain complete.

You will need

card or thick paper	scissors/a craft knife
wire	glue and sticky tape
batteries	bulbs and bulbholders
crocodile clips/paper clips	

1 You are going to make a string of buoys like the ones used to mark out shipping lanes. For each buoy you will need to cut out the shapes you see below from thin card: a semicircle for the body, a strip with windows for the lantern and a circle with a slit in it for the cone-shaped top. Use a craft knife to cut out the windows.

2 Assemble the buoys as shown below, fitting a bulbholder firmly into the body of each buoy with sticky tape.

3 Wire up the series circuit as shown on the left-hand side of the opposite page. Take a wire from bulbholder to bulbholder, completing the circuit from the last buoy back to the battery. You could put a switch in this section if you wish to.

4 Wire up the parallel circuit as shown below on the right. Take two wires, one from each battery terminal, to the two terminals on the first bulbholder. Then connect that bulbholder to the next one with two more wires. Add any other bulbholders to the circuit in the same way.

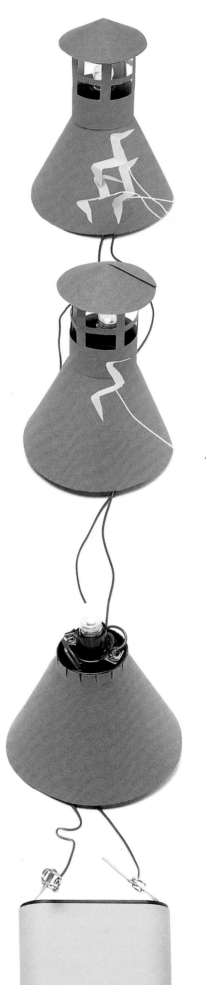

series circuit

parallel circuit

▲ Try taking the middle bulb out of each circuit and watch what happens.

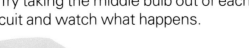

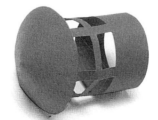

16 Light Bulbs

Light bulbs are used to produce light from electricity. The bulb contains a thin metal thread called a **filament**. When an electric current forces its way through this thin part of the circuit, the filament glows a bright white colour and the bulb gives off light.

You will need
a collection of old light bulbs
a craft knife
glue or tape
card
a ruler

opal globe bulb (100 W)   clear globe bulb (100 W)   photographic lamp bulb (600 W)

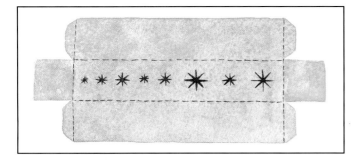

MAKE it WORK!

Light bulbs come in all shapes and sizes. Make a collection of different light bulbs, with a special box to display and store your collection. New bulbs are expensive, so just collect those which do not work any more. Handle light bulbs carefully, as the glass is delicate.

1 Work out what size you want your box. Then cut out a flat shape like the one on the left.

2 Fold up the sides of the box. Tuck in the corner flaps and glue or tape them in place.

3 Carefully cut the card along the top of the box so that you make spaces where you can stick the ends of the bulbs. Make sure that the bulbs fit in firmly.

4 Try to label your collection. Mark the different types of bulb (whether they have a filament or a fluorescent tube) and also how bright they are. You can tell the brightness of a bulb from the number of watts (W) of power that it uses.

opaque neon tube

The light bulb was invented in 1879 by an American called Thomas Edison. His first bulb used a piece of scorched thread as a filament.

car headlamp bulb

spotlight bulb

bicycle lamp bulbs

candelabra bulbs

professional photographer's flashbulb

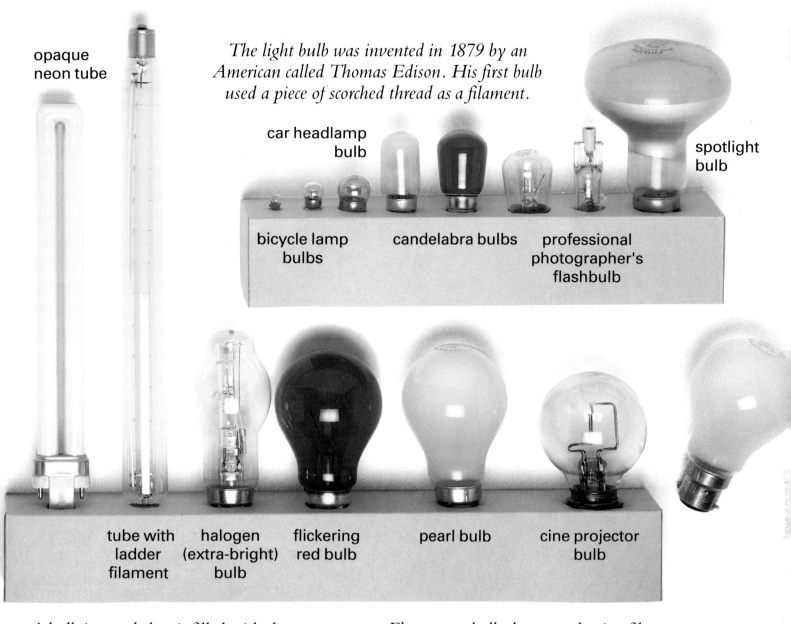

tube with ladder filament

halogen (extra-bright) bulb

flickering red bulb

pearl bulb

cine projector bulb

A bulb is a sealed unit filled with the gases nitrogen or argon. It contains no oxygen, the gas in the air that substances need to burn. Nitrogen or argon lets the filament glow, but doesn't allow it to catch fire.

Fluorescent bulbs have no glowing filament. Electricity is passed through a gas contained under pressure in the   bulb. The gas gives off light, but the bulbs   don't get very hot.

small clear globe bulb

small flickering bulbs

small spotlight bulb

18 Circuit Game

An electric current must always flow through a complete circuit. No current can flow in a broken circuit, because electrons have to keep moving in a continuous stream.

MAKE it WORK!

Test how steady your hand is with this circuit game. At the start of the game, the circuit is broken, so the light is off. If your hand shakes as you move the playing stick along, the loop touches the wire, the circuit is completed and the bulb lights up!

You will need

a battery	wire
a bulb and bulbholder	dowel
crocodile clips/paper clips	coloured tape
an old wire coathanger	
screw eyes, large and small	
balsa wood and wood glue, or a shoe box	

Most coathangers are lacquered with a thin layer of clear plastic to stop them marking cloth. You should rub away the plastic with a piece of sandpaper – otherwise the plastic will insulate the wire and the circuit won't work.

1 Make the playing stick by screwing a large screw eye into the end of the dowel. Connect a long piece of wire to the eye, and tape it down the length of the playing stick.

2 Make a box by gluing together pieces of wood, or use a shoe box. Paint the box, then divide it into sections with coloured tape.

3 Position the bulbholder at one end of the box, wire it up and push the wires through the top of the box.

screw eye   dowel

wire

4 Twist a small screw eye onto each end of the box. If you are using a shoe box, you may have to tape them in place. Bend and twist the coathanger wire to make the top part of the game. Slip the eye on the handle over the wire. Then connect the ends of the coathanger wire to the screw eyes on the box.

5 Beneath the box, connect one of the wires from the bulbholder to the battery. Put a crocodile clip on the other bulbholder wire and attach it to one end of the bent coathanger.

6 Connect the wire from the playing stick to the free battery terminal. Now you are ready to play the game!

circuit diagram of the circuit game

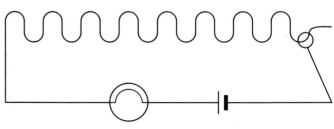

Playing the game

Hold the handle and move the loop along the wire without touching it. If your hand slips, the bulb will light up. Challenge your friends and see who can get furthest along the wire without the bulb lighting up.

▼ Make some different wire shapes to attach to your box. Try to make both easy and hard shapes. You will find that coathanger wire is quite stiff and you might need a pair of pliers to help you bend it, especially if you are attempting right angles.

Switches are used to turn electrical circuits on and off. When they are switched off, they break the circuit so that electricity cannot flow around it. When they are switched on, they complete the circuit, allowing the electricity to flow through.

MAKE it WORK!

Switches can be made to work in lots of different ways. For instance, you may not want a light to go out completely, but just to be a little less bright. Or you may need a switch that can turn a buzzer on and off very quickly, to make a special pattern of signals. Here are four different types of switch for you to try.

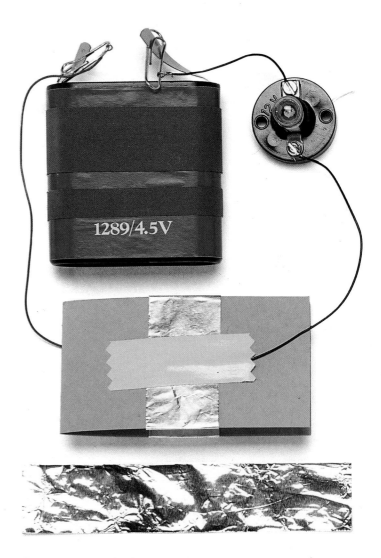

Simple switch

This is a simple on/off switch. When it is on, the current flows through the circuit; when it is off, the current stops. Wire up a simple circuit, like the one on page 8, but leave a break in the wires. Make a switch as shown above, using a block of balsa wood, a paper clip and two metal drawing pins. When the clip touches both drawing pins, the switch is on.

Pressure switch

This is the type of switch that can be used to make a doorbell ring when someone steps on a doormat. Wire up a circuit as before. Fold a piece of card in half. Wrap strips of foil around each half of the card, so that they touch when pressed together. Tape the wires to the foil on the outside of each side of the card. When the two strips of foil touch, the switch is on.

◄ These are low-voltage switches from a model shop. You can include them in any of the circuits shown in this book.

You will need

batteries	wire
bulbholders	bulbs
balsa wood	card
paper clips	tape
drawing pins	cork
aluminium foil	a pencil
strips of thin copper	a crocodile clip

Dimmer switch

Electricity can pass through the **graphite** in a pencil, but it is hard work. Graphite is called a **resistor,** because it offers resistance to the electric current. You can use a graphite pencil resistor to make a dimmer switch. The longer your pencil lead, the more resistance there is and the dimmer your light will be.

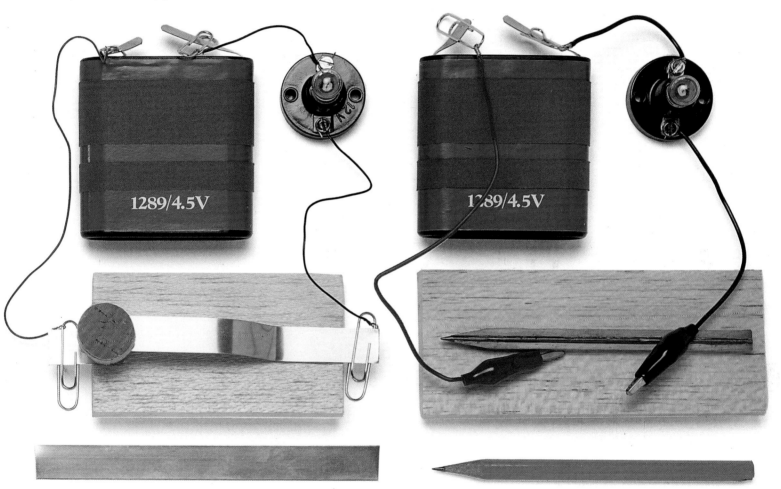

Tapper switch

This switch is used by Morse code operators. It gives the operator total control over the length of time the circuit is complete or broken. The switch is on when the two strips of copper are pressed together. It returns automatically to the 'off' position when not in use. The full instructions for how to make a Morse code tapper are given over the page.

Make a simple circuit as before, but fit crocodile clips to the free ends of the wire. Soak a lead pencil in water, and then ask an adult to slice it open down the middle. (**Be careful!** You should not try to cut the pencil yourself.) Attach the crocodile clips to opposite ends of the pencil lead, and then gradually slide one clip towards the other. What happens?

Morse code was invented in 1840 by the American painter and inventor, Samuel Morse. Each letter of the alphabet is represented by a simple combination of short and long electrical signals which can easily be transmitted down a single wire. The code is written down on paper as dots, dashes and spaces. Before the days of **communications satellites** and fax machines, all international newspaper reports and messages were sent flashing and buzzing down telegraph wires by Morse code operators.

MAKE it WORK!
Make a pair of Morse code tappers for sending and receiving secret messages.

You will need
two pieces of wood
two bulbs and bulbholders
two strips of copper
two slices of cork
two batteries
wire
paper clips
glue and screws

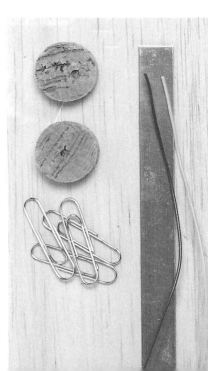

▼ International Morse code
These are the Morse code symbols. As you can see, they are made up of dots, dashes and spaces. A dot is transmitted by pressing and instantly releasing the transmitter key. To send a dash, hold the key down twice as long as you did for the dot. A space between letters is the same length as a dot, and a space between words is the same length as a dash.

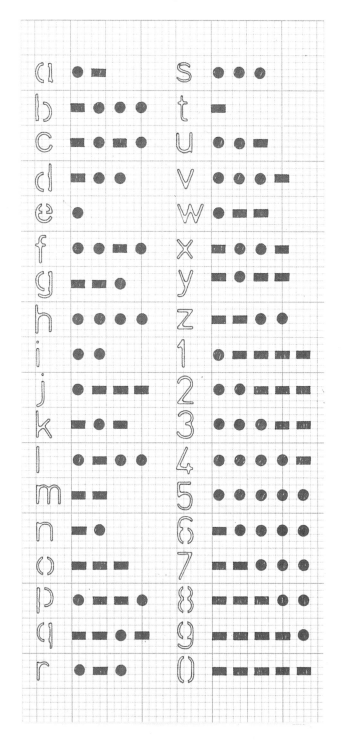

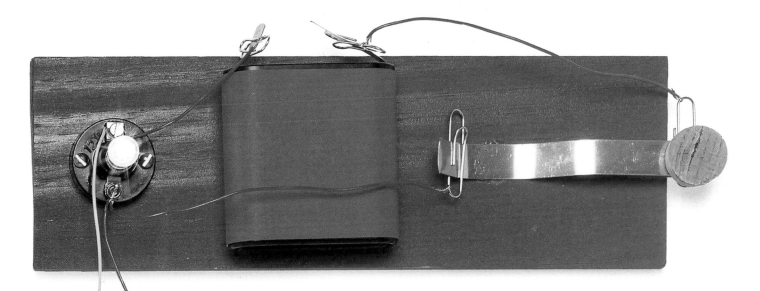

1 Glue the batteries to the boards and screw the bulbholders into position as shown.

2 Ask an adult to cut the copper strips into four pieces – two long and two short. Glue one small piece to the end of each board, making sure that they stick out a little. These are the bottom parts of the transmitter keys. To make the top parts of the keys, bend the long pieces of copper and fix them securely to the board, copying the shape in the diagram below.

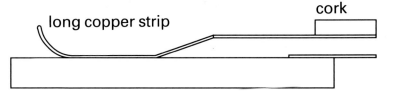

long copper strip

cork

3 Glue a slice of cork onto each transmitter key and wire up the keys as shown in the photograph. When you press the cork end, the long copper strip should touch the short one underneath and both bulbs will light up.

Some circuits are made up of lots of different connections, which act together to perform complex tasks. In electrical equipment such as radios, where many tiny circuits are needed, the circuits themselves may not be made as wires, but tiny strips of metal printed on a sheet. In computers, thousands of microscopic circuits are crammed onto one **silicon chip**.

MAKE it WORK!

Most circuits are made on a circuit board. All the wires are spaced out so they cannot accidentally touch one another. This question and answer game shows you what a simple circuit board looks like. Each connection, when correctly made, will complete a circuit and the bulb will light up.

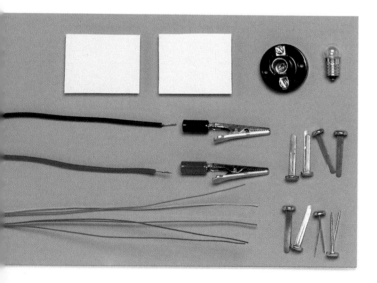

You will need

wire	card
a battery	paper fasteners
a bulb and bulbholder	a buzzer
stick-on Velcro tape	coloured pens
crocodile clips/paper clips	

1 Cut out a piece of card for your quiz board. Down each side of the card, push through a row of paper fasteners. On the front of the card, stick strips of Velcro backing tape next to each paper fastener.

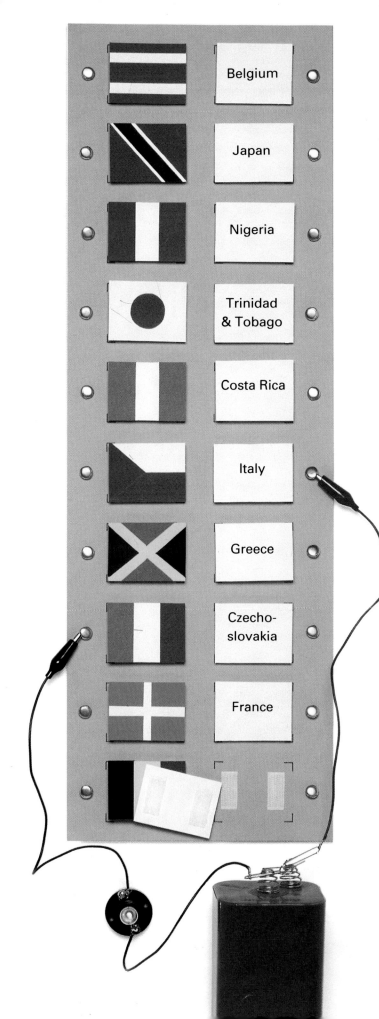

2 Make question and answer cards. Back them with Velcro and stick them down in random order. On the back of the board, wire up the questions to the correct answers.

3 Set up your testing kit of battery, bulb, wires and clips as shown.

4 Touch one of the paper fasteners on the question side with one of the testing wires. Then match it up with an answer on the other side. If you have picked the correct answer, the electrical circuit will be completed and the bulb will light up.

▲ Make raised shapes for your quiz board and replace the light bulb with a buzzer. Now you can play blindfold!

▲ ▼ Think up different quizzes for your board. What about animals, or tennis players?

Always check your connections carefully before you start to play. The game will not work if any of the wires become loose.

Just like static electricity, magnetism is a natural, invisible force. It was discovered over 2,000 years ago, when the Ancient Greeks first noticed that certain stones would jump together or move apart depending on which way they were facing.

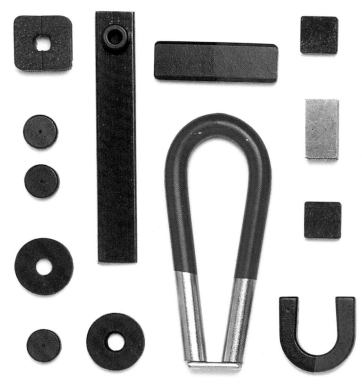

What is a magnet?

A magnet is a piece of iron or steel that attracts or repels certain other pieces of iron or steel. Like all other substances, metals are made up of the tiny particles we call **molecules** which, in turn, are made up of atoms. Normally, all the molecules in a piece of iron are facing in different directions. However, if we can rearrange the molecules and get them all facing the same way, they will act together as a magnet, making a powerful force.

MAKE it WORK!

You can watch the power of magnets at work with this fishing game. The object is to 'catch' as many high-scoring fish as possible. Players take it in turns to fish, and the winner is the player with the highest score.

1 Using the shape above, cut out large, medium and small versions of the fish from different colours of card.

2 Draw in the eye, gill and mouth using a thick black marker.

3 To get a fish-scale effect, stipple paint onto the fish using a stencil brush and a piece of wire mesh. Use a lighter coloured paint on the belly. Give each fish a score number.

You will need

thin card
thin wire mesh
a craft knife
a small magnet with a hole in the centre
paint and paint brushes
dowels and string
paper clips

4 Make fishing lines by tying the magnet to the string. Attach the other end of the string to a dowel rod as shown.

5 Attach paper clips to the fishes' noses.

6 Make a sea from a cardboard box covered in blue card. Put in the fish and start fishing.

Around a magnet is an area called a **force field**, where the pull or push of metal and magnet is at its strongest. The force field is strong enough to pass through wood or glass. The ends of a magnet, where most of the energy is directed, are much more powerful than the middle.

For the insects you will need

card
slices of cork
metal drawing pins

glue
paint and paint brushes
a horseshoe magnet

MAKE it WORK!

The more powerful the magnet, the larger and stronger its force field. See how a small magnet works through card, and how the force field of a strong steel magnet can even pass through a wooden door.

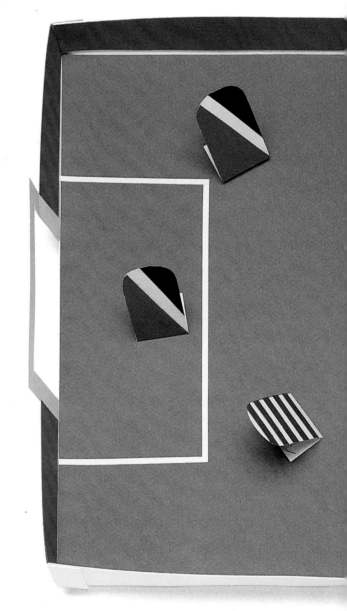

Magnetic insects

Using a craft knife, cut out insect shapes from thin card and paint them carefully. Stick each insect onto a small square of cork into which you have pushed a drawing pin. Use a strong horseshoe magnet to make the insects move from the other side of a door.

For the football game you will need

white card	green card
a craft knife	cork
metal drawing pins	dowels
small magnets	glue
a table football ball	paints or crayons

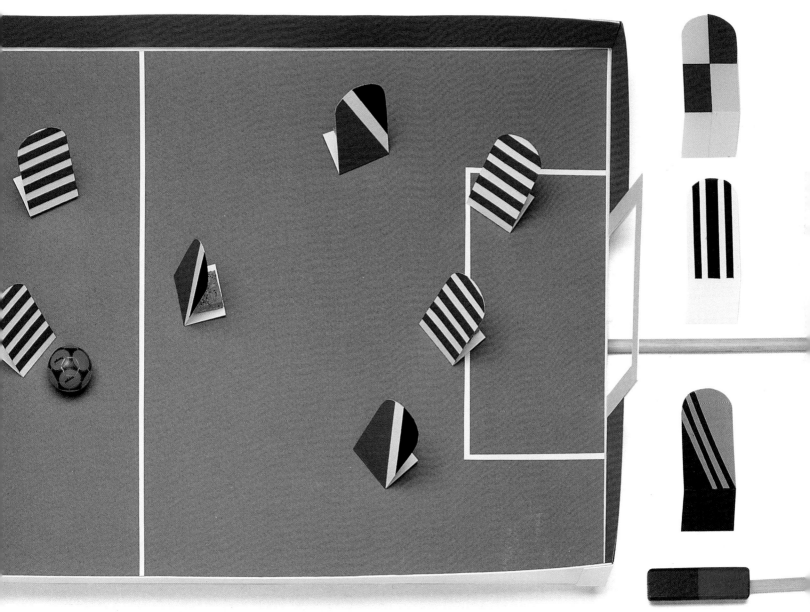

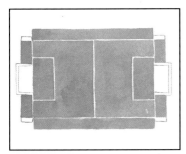

Table football

Make a box out of green card, folding and gluing the corners as shown. Mark out the lines of the football pitch in white. Make cardboard players with a piece of cork glued to the inside of each base. Stick a drawing pin through from the outside. The players are moved from under the pitch by magnets attached to dowels.

The force fields of magnets can pass through many different substances. The magnetic insects and magnetic football on the previous pages work because magnets can attract through wood and cardboard. A magnetic force field can also pass through water.

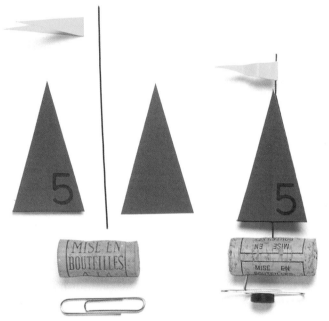

MAKE it WORK!

There are two different kinds of magnetic boat to make. The cork boats work by magnet to magnet attraction. The boat magnets are close to the bottom of the water container, so the boats can be pulled around by a small bar magnet attached to the end of a stick. The balsa-wood boat has drawing pins pushed into its keel and needs a stronger magnet with a force field that will attract through shallow water.

Even though the Ancient Greeks knew of magnets, for hundreds of years people did not know how to make magnets for themselves. It was not until the nineteenth century that magnetism, and its close connection to electricity, were properly understood.

You will need

thin, coloured card	wire
corks	paper clips
door magnets	glue
dowels	strong magnets
balsa wood	a wooden skewer
metal drawing pins	a glass tank

To make the cork boats

1 Make the sails out of coloured card. You can make one triangular sail by cutting out two triangles and sticking them back to back with the mast in the middle.

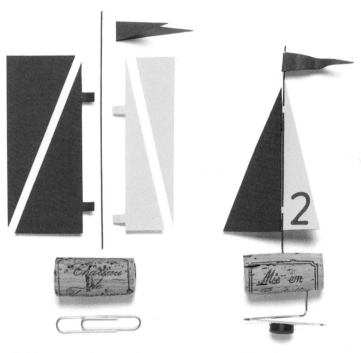

2 You can also make a more complex rig of a mainsail and jib. Cut a rectangle of card diagonally, leaving enough card on the straight edge to make two tabs. With these tabs, attach the sail to a piece of wire.

3 Push the wire mast into a cork. **Be very careful!** Do not stab yourself with the wire. Top the mast with a flag made from a folded strip of card of a contrasting colour.

4 Unbend a paper clip as shown above. Push one end into the underside of the cork and glue a small door magnet onto the other end, using waterproof glue.

To make the balsa-wood boat

1 Ask an adult to help you cut a deck and keel out of balsa wood as shown above.

2 Stick a wooden skewer into the centre of the deck to make a mast.

3 Make sails and flags as for the cork boats, but in a larger size.

4 Glue the boat together with waterproof glue and paint it. Push three drawing pins into the bottom of the keel.

To make the buoys

Put a short piece of wire into a cork and top it with a coloured flag. Stick a door magnet to the bottom of the cork with waterproof glue.

The Ancient Greeks had a mythical story about an island of magnetic mountains, which pulled the iron nails out of passing ships!

The Earth itself is a giant magnet and, like any other magnet, its strongest points are at the North Pole and the South Pole. No one really knows why the Earth is a magnet, but its force field extends thousands of miles into space. Any magnet on Earth allowed to swing freely will always point to the north – which is very handy if you need to find out where you are.

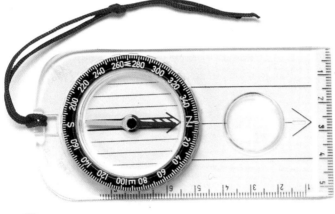

MAKE it WORK!

Because magnets always line up with magnet Earth, the end of a magnet that points north is called the north pole and the end that points to the south is called the south pole. Compasses are used to find directions, and they come in many varieties. Some are marked off with all 180 degrees of a circle. These compasses give a very accurate reading, but even the simplest compass will let you know if you are heading in the right direction. All you need is a magnetized needle that can swing freely.

To make a water compass you will need

an old yoghurt pot	a magnet
a needle	a slice of cork
card	a protractor

1 Cut out a circle of card and make a hole in the centre just smaller than the diameter of the yoghurt pot.

2 Using a protractor, divide the circle into accurate quarters and mark on the four compass points: north, south, east and west.

3 Make the needle magnetic by stroking it with one end of a magnet about twenty times. Always stroke in the same direction. Tape the needle onto a thin slice of cork.

4 Fill the yoghurt pot with water and float the cork in it. When the needle has settled to the north, tape the ring of card onto the pot. Check your readings against a real compass.

The poles of magnets react to one another just like the two kinds of electric charge. Opposite poles attract – and like poles repel.

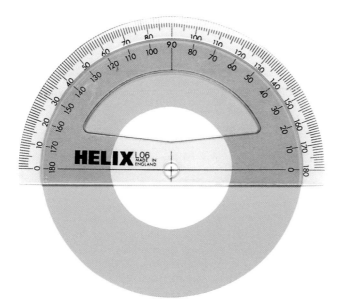

For two simple compasses you will need
card   needles
a magnet   tape

Fold a strip of card and tape a magnetized needle on it.

Jar compass
Suspend the compass strip in a glass jar using a straw or a pencil and some thread. This compass will work out of doors because the jar protects the needle from the wind.

◀ **Balancing compass**
Make a cone from a semicircle of card. Fix a wooden skewer or cocktail stick into the top of the cone and balance the compass strip on the end. This is strictly an indoor model!

You can actually see the force field that surrounds a magnet by sprinkling iron filings onto a piece of paper and then putting a magnet wrapped in paper down amongst them. The filings will rearrange themselves according to the magnet's force field, clustering around the north and south poles where the force is at its greatest.

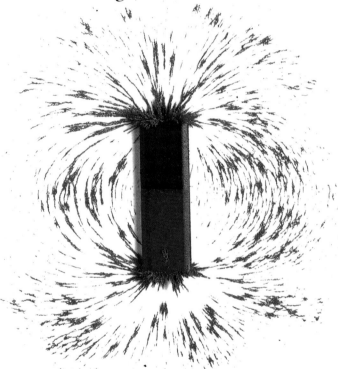

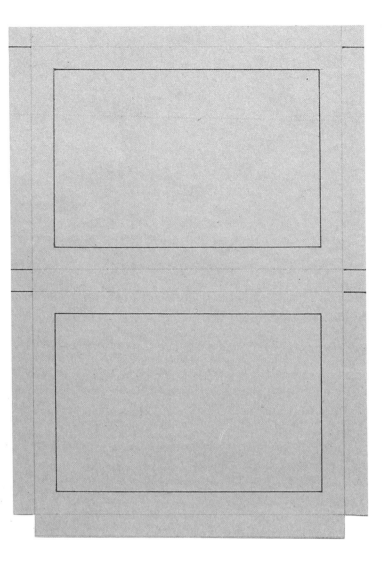

▲ Here you can see the force field at work. The iron filings form a pattern of lines running from pole to pole. These lines are called lines of force, and they show up the invisible force field of the magnet.

MAKE it WORK!
Use the power of magnetism to draw pictures with iron filings.

You will need

card	clear acetate sheets
rubber bands	iron filings and a magnet
scissors	clear sticky tape

Be careful! Iron filings are dangerous. Don't breathe them in or swallow any, and don't lick your fingers after touching them.

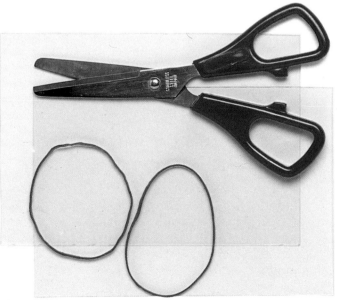

1 Cut out a rectangle of card and mark it as shown. Cut along the red lines and fold and glue along the pencil lines to make a box.

2 Take two pieces of clear acetate and cut them slightly bigger than the windows of the drawing box. Tape them in place with clear sticky tape.

3 Draw some faces on sheets of white card, leaving out the hair.

4 Put a face card inside the box and shake iron filings on top of the card. Snap the box shut with the elastic bands. Put the drawing box on a flat surface. Now you can 'draw' the hair on your face using a magnet.

A new kind of experimental train in Japan runs on the principle of magnetic levitation. Both the track and parts of the train are magnetic. It works on the pull and push of magnets that repel and attract. The train floats above the track because the train and the track repel one another. There are no wheels and tracks to wear out, and no **friction** *to slow the train down.*

We have seen that electricity and magnetism are closely related to each other. In fact, every electric current has its own magnetic field. This magnetic force in electricity is very handy. We can use electricity to make powerful **electromagnets** that can be turned on and off at the flick of a switch.

MAKE it WORK!

This crane uses an electromagnetic coil. The magnetic field produced by a single wire is not very strong, but when electricity flows through a wire coiled around a nail, the coil becomes a powerful magnet.

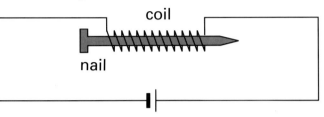

coil

nail

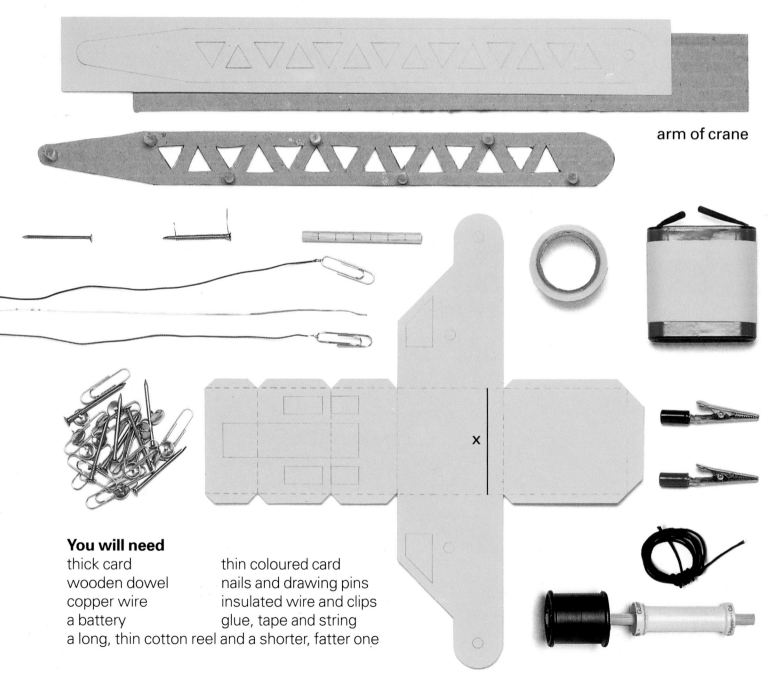

arm of crane

You will need

thick card	thin coloured card
wooden dowel	nails and drawing pins
copper wire	insulated wire and clips
a battery	glue, tape and string
a long, thin cotton reel and a shorter, fatter one	

1 Draw the shapes shown below onto the thick and thin card. Make sure that the line marked 'x' is the same length as the long, thin cotton reel.

2 Cut along the solid lines and fold along the dotted ones to assemble the body of the crane. Glue the thick card inside the body structure to make it stronger.

3 Put together the arm control winch as shown below, using the two cotton reels, dowels and a length of string.

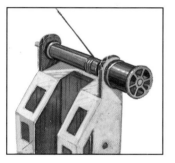

reinforcing card

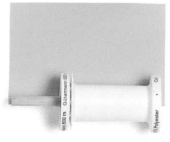

base support

reinforcing card

body of crane

4 Wrap copper wire around an iron nail to make an electromagnet. Then connect it to the battery with insulated wire, run over the top of the crane arm.

5 Tape the battery to the back of the crane. Check the circuit diagram to make sure the crane is correctly wired. When the leads and battery are connected the nail will become magnetized, and you can pick up a load of drawing pins. Disconnect and the pins will fall to the ground.

Electrical energy can be converted into mechanical energy, that is, energy that can pull and push and make things go. When electricity flows through the wires inside the motor, it makes them magnetic. The coil of wires becomes an electromagnet. It is attracted to fixed magnets inside the motor, which sets it off spinning around and around.

MAKE it WORK!

In this electric motor, a copper coil (the electromagnet) is connected to a battery by a clever little device called a **commutator**. The commutator brushes up against the wire leads of the electromagnetic coil, so that the electric current passes through; but the connections are loose enough to allow the coil to rotate freely.

As the coil turns, the connections of the commutator switch from side to side, so the direction of the electric current keeps changing. As the direction of the current changes, the poles of the electromagnet change sides too. The electromagnet is always attracted to the furthest fixed magnet and so it keeps on spinning and spinning.

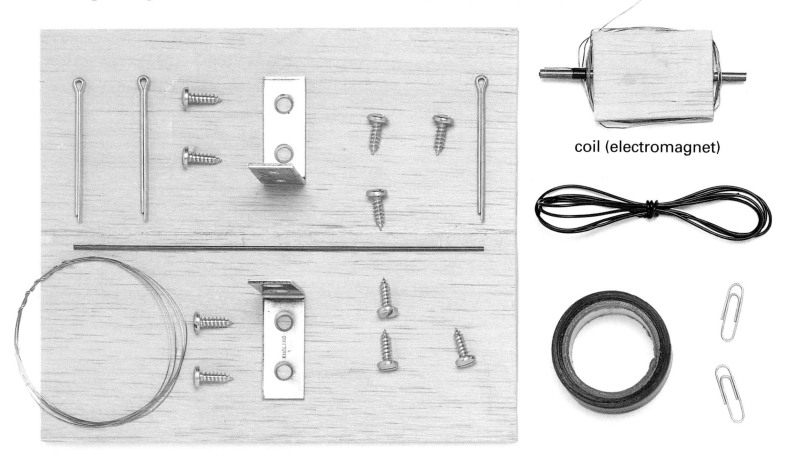

coil (electromagnet)

You will need

wood for the base	screws	copper wire	two strong magnets
two angle brackets	split-pin clips	balsa-wood block	insulating tape
thin copper tube	a metal spindle	crocodile clips/paper clips	a battery

1 Cut the base board out of a piece of balsa wood, or find a piece of soft wood. You could paint it a bright colour.

2 Ask an adult to drill a hole through the length of a small block of balsa wood. The hole should be wide enough in diameter for the copper tube to fit through.

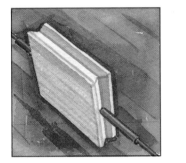

4 Screw the angle brackets to the base board. Stick on the magnets and position them so that they attract one another. Ask an adult to drill three holes along the centre of the board for the split pins to stand up in.

5 Thread the metal spindle through the split pins and the balsa-wood block so that the coil is suspended and can turn easily.

6 Now make the commutator. You have to get the wires from the battery to touch the ends of the wire from the electromagnetic coil without stopping the coil spinning round and round. You should strip some of the casing from the ends of the wires and bend them inwards. Follow the illustration on the left.

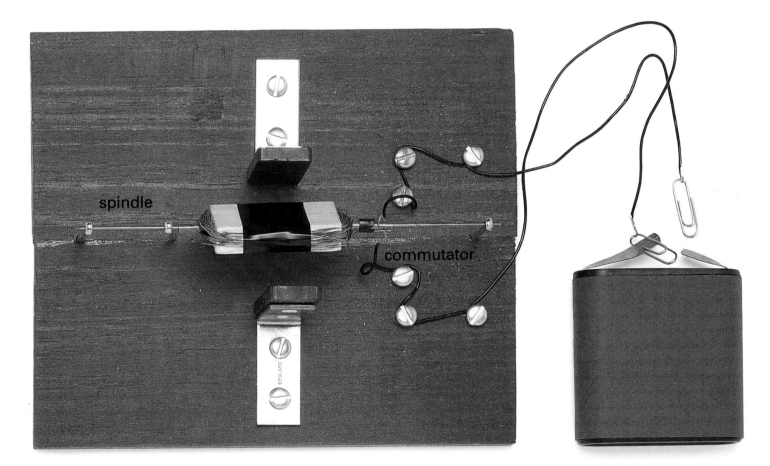

3 Cut grooves along two edges of the balsa-wood block and wrap copper wire tightly round the block. Insulate one end of the copper tube with clear sticky tape and fix the two ends of the coil wire in place with insulating tape as shown.

7 Screw the electrical wires into position so that the commutator connection is firmly fixed in place and cannot move. You may have to experiment a bit to get the screws positioned in the right place.

A simple electric motor turns a spindle round and round. One of the most direct and efficient ways of using this energy is to fix a propeller to the spindle. Boats are often driven by propellers in this way.

MAKE it WORK!

This propeller-driven boat makes good use of energy. It is designed with a propeller that drives through air rather than water because air is thinner than water and easier to move.

1 Make a balsa-wood framework for the hull and deck as shown on the right. Ask an adult to drill holes for the dowelling struts and glue them in place with a waterproof glue. Screw the electric motor in position. Paint the hull and deck with gloss paint.

2 Glue the battery onto the upper deck and wire up the circuit for the motor.

You will need

balsa wood	thin dowels
an electric motor	wire and clips
screws and nails	tape and glue
thin card	a propeller

To make the buoys

You can follow the instructions on page 31 to make marker buoys for your propeller boat.

3 Make the framework for the rudder, drilling shallow holes for the dowel to sit in so that it can turn freely from side to side. Glue and nail the balsa-wood support into position on the upper deck. Cut a rectangle of card for the rudder and tape it into place on the dowel as shown below.

4 Glue the propeller to the motor spindle, clip the leads onto the battery and watch your boat go! Turn the rudder to make it change direction.

Wrong way!

If your boat goes backwards instead of forwards, you may have fitted the propeller blades on the wrong way round, so that the propeller is pulling instead of pushing. The boat will also reverse if the battery is connected the wrong way round.

A hovercraft works in a similar way to this propeller-driven boat. It rides just above the surface of the water on a cushion of air, and is pushed forward by propellers positioned on top of the craft.

Modern high-speed electric trains are run by large electric motors. They take their power supply from overhead wires or electrified tracks. Not having to carry fuel increases their efficiency.

You will need

thin card	balsa wood
wooden dowelling	beads
an electric motor	paper clips
a battery	slices of cork
copper wire	upholstery pins
screws, thin nails and glue	
seven plastic bottle tops	
three thin copper strips	

1 Ask an adult to help you cut out the balsa-wood base of the engine. Then glue and nail the two roof supports into position as shown. Stick the upholstery pins into the supports.

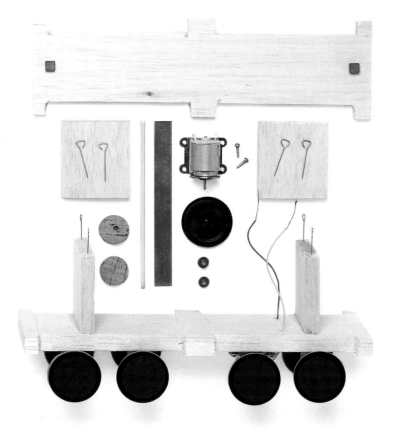

MAKE it WORK!

This model electric train works just like the real thing. The motor is on board but the power is supplied through power lines overhead and is passed down to the motor through the upholstery pins.

2 Glue a plastic bottle top to the spindle of the electric motor to make a wheel. Screw the motor to the base of the train, and then pass two wires from the motor up through the base to the upholstery pins.

3 Ask an adult to drill two holes into each copper strip and bend them as shown to make the axle holders. Drill two more holes and screw the axle holders into the base.

6 Lay down three long strips of card to make two grooves for your train to run along. Add balsa-wood pylons overhead and fix a length of wire between them, threading it through the upholstery pins. Connect the overhead wires to the battery and watch your train go!

Extra carriages
Experiment making different kinds of carriages for your train. They do not need motors or wires, but you can make wheels and bases in the same way as for the engine. Join them up with small door magnets behind cork buffers.

4 Make the axles by feeding the dowels through the axle holders. Then glue the beads and bottle tops to the ends of the axles to make the wheels.

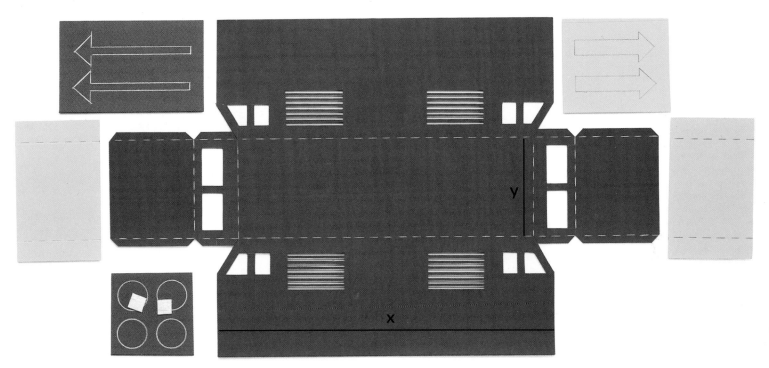

5 Mark out coloured card as shown, making sure that line 'x' is as long as the balsa-wood base and line 'y' is the same width. Assemble the body of the train and fit it over the base.

Running backwards and forwards
To make your train run the opposite way, just reverse the connections on the battery.

Electric motors range from the very small to the enormous. There are small battery-operated motors in model trains and clocks; but some electric motors in factories need a power supply so strong that it has to come directly from the power station.

MAKE it WORK!

See for yourself what happens if you double the power supply to an electric motor. Build a Spin-o-Matico to make colourful patterns with paint. Compare the different results you get with one battery and with two.

1 Glue the two tubs together, bottom to bottom, so you have two open ends. The top tub will be the paint tub and the bottom one will hold the motor and batteries.

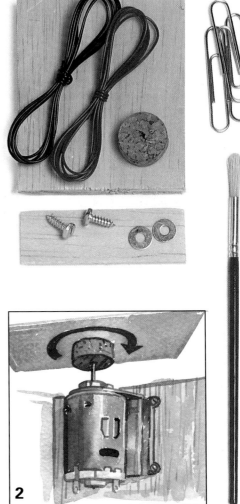

2 Poke the spindle of the electric motor through the centre of the bottom tub (the motor tub) so it pokes up through the bottom of the top tub (the paint tub). Screw the electric motor down onto a small piece of wood to hold it in place, and glue the wood into position.

You will need

two old margarine tubs	an electric motor
two 4.5-volt batteries	paper clips
screws and washers	wire
a slice of cork	wood
poster paint	card

3 Glue the batteries firmly to the bottom of the motor tub as shown. Then connect the two positive terminals and the two negative terminals with paper clips.

4 Make an on/off switch like the one shown on page 20, using a paper clip and two screws. Wire the batteries up to the motor, including the on/off switch in the circuit. Fix the on/off switch to the outside of the margarine tub, strengthening it on the back with a small piece of balsa wood.

5 Turn the margarine tubs over. Put a slice of cork over the top of the electric spindle that is poking up through the bottom of the top tub. Test to see if the motor is working and the cork whizzes around when you switch on.

6 Take a piece of card and stick it onto the cork using a dab of rubber glue.

7 Switch on the motor and dribble paint onto the whirling card to make a pattern. You can also use a paint brush if you want to.

full power

half power

▲ Try doing some paintings using half the power. Move the paper clips so they touch the terminals of just one of the batteries instead of both of them. How does the reduced power affect your finished painting?

When something is spinning very fast, the outside spins much faster than the inside. The force that seems to push everything towards the outside is called centrifugal force.

Acid A certain kind of chemical. Foods that contain acids, such as lemons, taste sour or sharp. Very strong acids are dangerous, and can burn holes in wood or cloth.

Alkaline A word used by chemists to describe a chemical property of certain substances. In chemistry, an alkali is the opposite of an acid. One common kind of alkali is magnesia, the white liquid or powder we take to cure an upset stomach. Caustic soda is an example of a very strong, dangerous alkali.

Atomic power Energy that comes from making changes to the centre of an atom. By splitting atoms, an enormous amount of heat is created. This heat is used to boil water, making steam to drive turbines and produce electricity.

Atoms Tiny particles, over a million times smaller than the thickness of a human hair. Everything around us is made up of atoms – they are like building blocks, and by combining different atoms in different ways, different substances are created.

Circuit A loop-shaped path along which electricity can flow.

Communications satellite A spacecraft that goes round and round the Earth and is used to relay telephone conversations, fax messages and television pictures from one part of the globe to another.

Commutator A special kind of electrical connection, used in an electric motor. A commutator makes the direction of the electric current change at regular intervals.

Component In electronics, a component is one single part of a whole circuit. For example, a switch or a battery is a component in an electrical circuit.

Conductor In electronics, a conductor is any substance that an electric current can pass through.

Current electricity Current electricity is the electricity we use in homes, offices and factories. It is produced in power stations, and is then distributed around the country through wires, pylons and transformers.

Electrolyte A liquid solution that is able to conduct electricity. Batteries use electrolytes to make electricity.

Electromagnets When an electric current passes through a metal, such as a piece of iron or copper, it always produces a magnetic field. Electromagnets are especially useful because their magnetism can be switched on and off with the electric current.

Electrons Tiny particles of atoms. Each electron carries an electrical charge.

Electroscope A scientific instrument used to measure the strength of an electrical charge.

Energy Energy is needed to do any kind of job or action. Motors and engines use energy, and our bodies do too. The food we eat, or the electricity that powers an electric motor, are called energy sources.

Experiments Scientists do experiments to test out their theories about how the world works.

Filament A thin coil of wire, usually made from a substance called tungsten, inside a lightbulb. The electricity has to work so hard to push its way through the tungsten that the coil glows and gives off light.

Force field The area around a source of energy (such as a magnet) where the energy works.

Friction Friction happens when one object moves side by side to another one. Friction produces heat (like when you rub yourself to keep warm) and makes the two objects stick together (like tyres gripping the road).

Generator A machine that turns heat or movement into an electrical current.

Graphite The substance that pencil leads are made out of. Pencils used to contain actual lead until people discovered that graphite was better for writing with.

Hydroelectricity Electricity that is produced by the movement of water through a generator.

Insulators Materials that do not conduct electricity. Rubber and plastic are both good insulators.

Load The part of an electric circuit that uses the electric power. In a lighting circuit, the load is the light bulb.

Matter All the different substances in the universe are matter. It comes in three forms: solids, liquids or gases.

Molecule A tiny particle of a substance. Every molecule is made up of two or more atoms joined together.

Physics The study of energy and matter.

Resistor A substance which offers resistance to an electric current. The current has to work very hard to get through a resistor, so resistors are put into circuits to reduce the voltage of the current.

Scientist Someone who studies the world in a systematic way, to try and understand how it works.

Silicon chip A tiny, wafer-thin slice of the substance silicon, which has a whole electronic circuit on it, in miniature. Silicon chips are an especially important part of modern computers.

Static electricity Static electricity is an electrical charge, produced naturally when two things rub together. Lightning is the best-known example of static electricity.

Terminals The points in an electrical circuit where the electric current leaves or enters the circuit.

Theory An idea which tries to explain something. Scientific theories usually have to be proved by experiments before they are said to be true.

Ancient Egyptians 13
Ancient Greeks 26, 30, 31
argon 17
atomic power 4
atoms 6, 26

batteries 5, 10-11
battery tester 11
boosting power 44-5

centrifugal force 45
circuits 8-9, 12, 14-19, 20
coin battery 11
commutator 38, 39
compasses 32, 33
computers 24
conductors 8, 9

Edison, Thomas 17
electric motor 5, 38-44
electric trains 42-3
electrodes 10
electromagnetism 36-7, 38-9
electrons 6, 7, 8, 9, 18, 28
electroscope 6

filament 16, 17
fluorescent bulbs 17
force field 28, 30, 32, 34

generators 12
graphite 21

hovercraft 41
hydroelectricity 4, 12

insulators 8
iron filings 34

light bulbs 5, 16-17
lighthouses 12-13
lightning 7
lines of force 34
load 9

magnetic fields *see* force fields
magnets, magnetism 5, 26-35, 38-9
molecules 26
Morse, Samuel 22
Morse code 21, 22-3

negative charge 6-7, 10
neon bulbs 17
nitrogen 17
nuclear fuel 12

parallel circuit 14
Pharos of Alexandria 13
physics 4
Poles (of the Earth) 32
poles (of magnets) 32, 34, 38
positive charge 6-7, 10
power stations 12, 44

radios 24
resistance, resistors 21

series circuits 14
silicon chips 24
static electricity 6-7, 8, 26
switches 20-21

terminals 8

Volta, Count Alessandro 11
volts 13

watts 13, 16-17

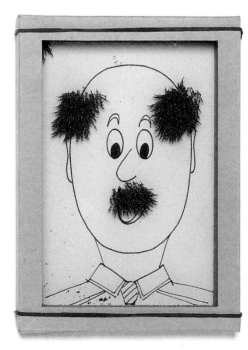